# SHE'S WATCHING

*The Revelation for the Black Church
and Child Sexual Abuse*

by

*Stephanie M. Myers-Lewis, DMin*

Watersprings
PUBLISHING

SHE'S WATCHING: *The Revelation for the Black Church and Child Sexual Abuse* published by Watersprings Publishing, P.O. Box 1284 Olive Branch, MS 38654
www.waterspringspublishing.com

Contact publisher for bulk orders and permission requests.

Printed in the United States of America.

ISBN-13: 979-8-9859594-0-6

# Table of Contents

*Behold, children are a heritage from the Lord,*
*the fruit of the womb a reward.*

*- Psalm 127:3*

# Acknowledgments

This work would not have been possible without the support of so many individuals, including my mother, Shirley V. Myers, who believed me and in me. And, to my late Pastor, Rev. Wardell Johnson, who was placed in a teenager's life when she desperately needed the Word of God like the air she breathed, I am eternally grateful for you both.

Thanks to my St. Mark Baptist Church family and Pastor Keith Lashun Mcgee, who "saw me and the gift I carried" when I didn't want to be seen. My ministry has been enriched by your pastoral heart and demands to utilize the gift God has entrusted to me. I am grateful for your ministry and my church family.

I would like to thank my St. Paul Baptist Church family, Rev. Dr. Pastor Christopher B. Davis, and Executive Pastor Mary E. Moore for creating a space for my women's ministry gift to grow while demanding excellence and accountability in my leadership. Thank you to my St. Paul Young Adult Class, Women's Ministry colleagues, Memphis Pastors & Clergy Leaders, Memphis Child Advocacy Center, and the many women and men who were brave enough to entrust me with their stories.

I am grateful to those who granted me the opportunity to conduct one-on-one confidential interviews. I am most appreciative of those who granted permission to record sessions as a means of pondering and gathering information needed for this project. I am grateful to each of the members of my

Dissertation Committee. Thanks to all my Memphis Theological Professors and colleagues in the ministry. I am richly indebted to you all.

No one has been more important to me than my family members, children, and my sisters in Christ, who continually encouraged me. A special thank you to my friends Terri Starks, Jacqueline Hayes-Fennell, and Sisters in the Ministry Pastor Ella Mosby, Rev. Marilyn Smith, Dr. Felicia Lavant, Rev. Veverly M. Edwards, and Dr. Jeralyn B. Major, for always encouraging me and keeping me lifted in prayer, and to my greatest supporter in all things, Mr. Lafayette Lewis. You all, most especially my family, are my ultimate supporters who provided never-ending encouragement to propel me toward completing this project.

*"To be called beloved is not only to shatter the silence, but to get rid of it altogether."*

Emilie M. Townes

*"The only thing necessary for the triumph of evil is for good men to do nothing."*

*- Edmund Burke*

# Introduction

The purpose of this book is to confront the silence surrounding child sexual abuse within the Black Church. When the psalmist penned these words from King Solomon in Psalm 127:3-4 (NLT), *"Children are a gift from the Lord; they are a reward from him,"* the Israelites understood the essence of the meaning as a song of worship that they would sing on their pilgrimage to the Holy City. As they ascended to Jerusalem, their worship was centered on reverencing God and acknowledging the inheritance that children are as sacred gifts from God.

The musicality of this song was one in which the Israelites would elevate their praise to God by repeating the words in the hymn. From a scholarly perspective, there is an inherent belief within this repeated refrain that there is an expectation that God's children will be treated as gifts. When this expectation is not heeded, and instead there is a breach in safeguarding our children — especially in the church — the impact and trauma are inextricably tied to what happens in the worship atmosphere.

*It is critically important that pastors and churches within the African American community unify to confront its insidious culture of "Protectionism."*

# CHAPTER 1

# Child Sexual Abuse and the Black Church

The issue of Child Sexual Abuse (CSA) continues to wreak havoc upon the lives and future of our most vulnerable asset – our children. This social issue transcends political, racial, gender, socio-economics, and denominational constructs. Yet, there is arguably no social issue that is more egregious and can traumatically impact our children's lives.

According to national child abuse statistics from the Childhelp Organization, "Our children are suffering from a hidden epidemic of child abuse and neglect. It's a widespread war against children that we have the power to stop, and understanding the issue is the first step." The long-term effects of child sexual abuse are staggering, from shortened life expectancies to the development of psychological disorders, even to devastating financial impacts estimated upwards of $100 billion in lost worker productivity, healthcare costs, education costs, and criminal justice expenditures, to name a few.

According to Rights4Girls, a human rights organization dedicated to ending gender-based violence against young women, until our communities understand the urgency of addressing this issue, then the work of dismantling the systems that create and perpetuate the pathway for sexually abused girls into the prison system will remain a burden that is on the shoulders of a few. "Research reveals that girls sent into the juvenile justice system have typically experienced overwhelmingly high rates of sexual violence," especially little Black girls. Even more telling is when the Black Church is silent and abdicates its responsibility of being a prophetic voice for the broken and marginalized, who, because of traumatic experiences, are often unable to amplify their voices.

As reports of CSA continue to emerge, it is possible that relatives, friends, colleagues, and even church leaders will be among the accused. Many studies demonstrate that faith communities are more vulnerable to abuse than secular environments. "The Abel and Harlow study revealed that 93% of sex offenders describe themselves as religious. Other studies have found that sexual abusers within faith communities have more victims and younger victims. The disturbing truth is perhaps best illustrated by the words of a convicted child molester who told Dr. Salter, 'I considered church people easy to fool…they have a trust that comes from being Christians. They tend to be better folks all around and seem to want to believe in the good that exists in people.'" (https://religionnews.com/2014/01/09/startling-statistics/)

Child molesters and sexual predators have escaped the attention of the Black Church for so long. So it is no wonder they feel invincible in assaulting multiple victims. Oftentimes our Black Churches struggle to deal with the abuse of black girls and women because addressing this widespread problem ultimately means singling out the black male. Unfortunately, many are hesitant to do this because they fear being another vehicle that contributes to destroying the black man.

While this line of thinking is admirable and has been deemed necessary as part of the Black Church culture for the many who earnestly seek repentance, this cannot be an acceptable practice for sexual trauma by the black male that is rarely acknowledged and that renders the victim silent. Holding our Black clergy accountable for the culture they created, whether by commission or omission, is crucial for the survival of our children. Whether they are the victim or perpetrator, our little Black boys and little Black girls are gravely impacted by the continual silence.

Acquiring national research data or statistics specifically targeting CSA reporting within the Black Church and community was problematic due to the lack of national or statewide reporting being readily identified. If the Black Church were to conduct such a national study and investigate reporting from police precincts, its numbers would likely parallel the national CSA research reporting in other non-Black congregations.

These cases, and the many reported and unreported cases that may never be discovered, should signal today's Black preaching voice to delve deeper into exposing the issues that enable CSA. "Without a centralized theological body, evangelical policies and cultures vary radically, and while some church leaders have worked to prevent abuse, many have not. The causes are manifold: authoritarian leadership, twisted theology, institutional protection, obliviousness about the problem, and perhaps most shocking, a diminishment of the trauma sexual abuse creates—especially surprising in a church culture that believes strongly in the sanctity of sex." (https://www.washingtonpost.com/news/posteverything/)

Edmund Burke said, "The only thing necessary for the triumph of evil is for good men to do nothing." Patriarchy is not exclusive to the culture of the Catholic Church, Southern Baptist Church, or the Protestant Church. There cannot be an earnest call for the victims to speak out without the voice of the Black pastor and Black clergy leaders. Because the victims' voices have been muffled or silenced for so long, any chance of eliminating this pattern of abuse requires revisiting structural and theological truths in Black preaching.

It requires a willingness to expect criticism, rejection, and anger from the voices that deny, ignore, or fear Black preaching that admonishes the perpetrator. "The most important question for this discussion is, how will the church respond to this movement? How will the Black Church respond to this movement? What will we preach? Will we push deeper and deeper into these bigger subjects with real honesty and search for truth?" (Thomas, Frank A., *Introduction to the Practice of African American Preaching*. Abingdon Press. Kindle Edition.)

Not only are our little Black boys' and little Black girls' bodies being slaughtered by the complicity of silence, but their faith in God is also being crucified. Our lives revolve around an everchanging world that oftentimes mirrors a reflection of patriarchal culture indoctrinated into religion as absolute truth. The Black Church, at times, falls back on its adopted cultural and social assumptions in a field dominated by the male gender. The underrepresentation of female clergy leaders within the body of the Black Church is apt to skew the belief that biblical doctrine is perfectly objective and advocates issues without bias. However, the gender of the theologian in theory and practice can matter when addressing taboo issues that tend to affect one gender greater than the other. The Bible is explicitly recorded, interpreted, and preached from a patriarchal lens that has silenced and left a void for the voice of girls and women.

Professor of the New Testament Dr. Mitzi Minor continues to challenge seminary students to "study the Bible with a hermeneutic of suspicion, always being mindful of the voices that are not being heard." The usage of gender-neutral language is slowly being accepted in some seminaries and a few Black Church Pulpits. However, the dominant language is one of patriarchy and remains the accepted norm in most congregations. With the issue of child sexual abuse so widespread and impacting girls more than boys, can the gender of the preacher impede a prophetic message that could revolutionize how we preach to our children and their perpetrators?

As a daughter of the Black Church, a female preacher, a lover of the Black Church family, and a survivor of child sexual abuse and molestation, I own my right to critique perpetually that which I love. While the Black Church has very well been and shall always be a source of hope and healing, it needs reform. The Black Church pulpit has always been known for being the strong voice for justice, morality, ethical responsibility, spiritual renewal, and redemption. It is most disheartening that Black Preaching does not sound the alarm to admonish the predator and would-be predators that continue to rape, molest, and traumatize our little Black girls in record numbers.

What would Black Church look like were it to embrace equity in its theology? Gender equity in its social justice rhetoric? And equity in its liberation preaching? Does Black Preaching bear responsibility for preaching against the incessant abuse of our little Black girls within our churches and community at large? This line of questioning and exploration of Black Preaching advocacy for our little girls in no way suggests that boys are not being sexually violated or molested. This book confronts, identifies, and challenges Black Preaching to shift its theological, cultural, and societal constructs to one that advocates for our largest community of victims, which statistically identify as little Black girls.

*"If anyone causes one of these little ones—those who believe in me—to stumble, it would be better for them to have a large millstone hung around their neck and to be drowned in the depths of the sea."*

*- Matthew 18:6 (NIV)*

# CHAPTER 2

# Resistance to Change

Having engaged in many conversations with family members, friends, and clergy leaders in the African American community throughout this project, I was most disheartened to witness the apprehension of those who remain offended by the voices of liberation and refusal to see or hear the cries of the oppressed.

Michelle Rhett, an attorney and national advocate for children, suggests, "If the human stain of America is racism, the spiritual stain of the Church is sexual misconduct. The Church is a neon sign for perpetrators of sexual misconduct." Somewhere a child is sitting in the pews on a Sunday morning or Wednesday night, wondering like John did when he was sent to prison for preaching the gospel if this is the Church, the sacred, liberating venerable institution for all, including children? The Black Church culture has been no less guilty than any other institution as it relates to its sinful silence toward child sexual abuse.

During my research, I began to read stories about the lives of some of the most high-profile reports of CSA. For example, Bishop Eddie Long was deemed one of the most respected and charismatic gifted preachers in the Black Church. Yet, he was accused of grooming and molesting little boys. As recent as 2017, three Ohio pastors were indicted on sex trafficking charges, and prosecutors say all three men worked together to entice underage girls with money in exchange for sex.

While conversating with other clergies in the Black Church, I discovered articles reported by numerous sources that C.L. Franklin, "the pastor with the golden voice," had molested a little Black girl. I found not one article that disproved or confirmed these reports to be false. Even so, Franklin became the pastor of another Black congregation that welcomed him with open arms.

Why? Because it was the collective thought that he was considered a role model for the voices of Black Preaching. His gift for preaching and his classic sermons excused his sexual misconduct. This perverted way of celebrating in the Black Church is no different than the child sexual predators in Hollywood with legendary musical geniuses. Black Pastors have a history of covering egregious acts with biblical passages such as Romans 3:23, *"All have sinned and fallen short of the glory of God."* While this is a biblical truth, it has been misused by many pastors and preachers to protect the predator and, in doing so, shamed the victim into silence.

While I was teaching a teen class on a Sunday morning, a 14-year-old boy declared that he wanted to be a pimp like his daddy when he grew up. Thinking this black male teen was being sarcastic, I continued to ask questions. I turned to glimpse the look at the "How dare you" expressions on the faces of the teen girls present. Upon realizing his seriousness and irritation with my line of questioning, I asked him if he thought that any of the young girls who were present would allow him to, as he stated, "pimp them"? He responded, "They may not let me, but I know a lot of girls who will." This is our current context, and our children deserve more attention than the present state.

It is critically important that pastors and churches within the African American community unify to confront its insidious culture of "Protectionism." This term has been used to cover or excuse the sexually deviant behavior of the authoritative, charismatic, and gifted Black male leader. It continues to enable the incessant sick behavior that is now being called out in churches and communities abroad. The issue of the Black Church's silence is not indicative of willful intent on the part of early mothers and fathers of the Black Church; rather, it is merely a reflection of the patriarchal culture that created an oppressive indoctrination in the life of the Church.

"Prior to the 1980s, the sexual abuse of children and adolescents was a well-kept secret throughout society. The women's movement of the 1970s dragged incest and sexual abuse into the public eye. Throughout the 1980s and 1990s, methodologically sound empirical studies were published indicating that approximately one-third of women and one-fourth of all men are sexually abused prior to the age of eighteen." (Frawley. Kindle Edition) According to the national statistic on child sexual abuse, these numbers only reflect reported cases. As were then and now, most often, these children were abused by someone they knew and trusted, someone who had authority over them.

While there have been many debates surrounding child sexual abuse and tradition, the deafening silence from the Church as it relates to the urgency in addressing victims of child sexual abuse is concerning. I am unequivocally convinced that the "#MeToo" movement demands that the Black Church extend its voice to the fight against CSA. This movement, a collective echo that has given license for the abused to amplify their voices, has burst the damn of the silence of its victims. The many survivors, who range in age from thirties to sixties, interviewed during research for this project were still struggling with PTSD — a telling sign of the long-lasting impact that abuse has — since the abuse occurred many years ago when they were children. And according to recent statistics, the high incidence of child sexual abuse inside and outside the Church shows no sign of decreasing.

# CHAPTER 3

# Need for Change

The theological catalyst for this book is embedded in the book of 2 Samuel, which depicts the rape of Tamar and will be further explained in the proceeding chapter. Tamar's story encompasses the many characteristics of church, family, and community that continue to enable CSA toward little Black girls. Jesus warns against the exploitation and abuse of our children, inferring that the gospel is holistic and should address every area of our lives.

To miss the teaching and preaching to our children regarding the pervasiveness and prevalence of sexual abuse is sinful. Sin in its Greek meaning, ἁμαρτάνω, denotes "missing the mark." To model this Christocentric teaching, it should be incumbent upon the Black Church to embrace a holistic model that teaches our children not to be afraid to reject and report the sexual predator, no matter who it is, without fear of retribution. Jesus intentionally lifts God's love for children in Matthew 19:14 (KJV), *"But Jesus said, Suffer little children, and forbid them not, to come unto me: for of such is the kingdom of heaven."* Maybe we are not intentionally forbidding the little ones, but we are hindering their pathway because we are silent about what matters. Black Preaching has a responsibility to advocate for our "little ones."

Jesus further asserts in Matthew 18:6 (NIV), *"If anyone causes one of these little ones—those who believe in me—to stumble, it would be better for them to have a large millstone hung around their neck and to be drowned in the depths of the sea."* By missing the mark, our silence gravely participates in the stumbling of our children. "Some crimes against humanity are so heinous nothing will ever rectify them. All we can do is attempt to understand their causes and do everything in our power to prevent them from happening, to anyone, ever again." (Walker. Kindle edition)

The concept for this book was initially to challenge the Black Church to acknowledge, confront, and implement practices to minister to victims and survivors of the #Metoo movement. However, a shift manifested while engaging the community, interviewing clergy, and hearing the stories of (CSA) survivors' experiences within the Black Church. This shifting created an opportunity to explore preaching possibilities to confront the theological inerrancies and silence that continue to enable CSA within the Black Church and the Black community. Borrowing from James Cone parenthetically, "To speak of God and God's participation in the liberation of the oppressed of the land is a risky venture in any society. But suppose the society (Church) is racist (sexist) and uses God-language as an instrument to further the cause of human humiliation. In that case, the task of authentic theological speech is even more dangerous and difficult."

# CHAPTER 4

# Biblical, Theological, and Cultural Perspectives on Child Sexual Abuse

*"You cannot switch my gender and tell the same story. The parameters would all shift. I believe this simple truth lies at the heart of sexual harassment and assault. When gender makes women uniquely vulnerable, and inescapably inferior, the stage is set for victimization."*

Ruth Everhart

For the Black Church to admit that its theology and preaching enabled a fertile culture for sexual predators would demand repentance of its horrible sexual misconduct and abuse history. For example, while conversing with some project participants, a 32-year-old man blurted out, "They will not talk about child sexual abuse because they don't want to stop their own sexual behavior. It's too close for comfort." Whether this is true for the masses is another project. However, disregarding their responsibility to confront, identify, and change oppressive theology is a disservice to the Black Church and community.

The Bible is rich with examples of child sexual abuse and exploitation. To infer that child sexual abuse is a new problem or somehow a product of our current context is to disassociate the biblical narrative, culture, and tradition that the Black Church has always tolerated. The careful interpretation that gives voice to the dehumanization and devaluing of girls and women demands a lens of suspicion in studying and reflecting on the history of Black theology.

There are many texts and sermons that clearly ignore the sexual trauma of children in the Bible. Just as the ancient text silences the voices of girls and women, so does Black Preaching in many churches. Because Black Church preaching and teaching is male-dominated and promotes/supports male entitlement, there will always be the likelihood of dehumanization, devaluing, and an oppressive landscape that is outright dangerous for our Black girls. When a white person clings to the heritage of celebrating their heroes, who brutalized, raped, lynched, demonized, and degraded Black bodies with no redress of the trauma their heroes caused, we call it blatant racism. Many white people are insulted and angered at the mere suggestion that their memorials and altars to slave owners, rapists, and murderers are offensive and psychologically traumatizing to the descendants of slaves. The complicity of silence and indifference toward child sexual abuse is equally disturbing and comparable in effect on the survivor, the Black Church, and the community.

Pastors have a moral responsibility to be a prophetic voice in the community. It is an inherent burden centered on the call to preach a liberating gospel that is uncomfortable and unpopular. In the same way prophets in the sacred texts wrestled with the messages God imparted through them, the same rings true for preachers today. Failure to address and approach taboo issues through the prophetic gift of preaching is an affront to God, and it leaves little hope of healing for survivors of child sexual abuse.

The magnitude of Tamar's pain is evident in this passage. Tamar confides in her brother, Absalom, about the rape she has just experienced at the hands of her half-brother, Amnon. Absalom is mad, angry, and wants to kill him…But? A quick reading of the story and reminiscent ways of preaching this text would make one deem Absalom the hero. However, a careful and progressive suspicious reading of the text would reveal the motive for wanting to kill Amnon. It had little to do with fighting for the honor of his sister Tamar and everything to do with a devised scheme to use it as a pawn to inherit his father King David's crown" (West."Tamar").

The Black Church often does a disservice when it fails to preach this text through the lens of Tamar's pain, opting instead to treat her as an interloper who becomes collateral damage in the story of her brother Absalom's pursuit for the throne. The Black Church can no longer assign impunity against both Amnon and Absalom. Instead, Black Preaching must hold these two men responsible for their actions and use this passage to address the pervasiveness of today's rape culture against young girls.

Tamar knew the law, and she followed the law to cry out to God about her rape, yet nothing was done. In a culture dominated by patriarchal leadership, laws are made to be broken for those in power. Neither America nor the Black Church can legislate morality or gender equality. Were the laws, then and now, interpreted and rendered in a gender-neutral culture, perhaps Tamar's story and those who are victimized in the current culture would be different.

The voices of all the #MeToo survivors and victims are "crying out to the divine" like Tamar, with the expectation of the

Black Church to create spaces in the pulpit and conversations in the pews that respond to the cries of the oppressed. However, the dehumanization of female bodies will continue to be the norm lest our churches speak out and the conversation about CSA becomes a part of the pulpit and the Black Church.

It matters who is preaching to our children. It matters that gender-neutral language is utilized and that the preaching voices are diverse and inclusive of the female preacher. The expectation that our black preaching and culture should have changed since the biblical travesty of Tamar's rape is not fully evident in our context. In the Black Church, still, "Women are expected to sit in the pews, receiving messages from men in the pulpit. Their role is to recognize God in their pastor, not to expect or demand that he recognizes God in them." (Melissa Harris-Perry)

Many times, Black Preaching alienates the voices that would challenge normative prosperity-focused preaching, mainly female voices. There is seemingly little space for voices that speak a truth that is offensive, regardless of its relevancy to the life of the church and community. There is always the warning for female preachers to 'Just preach the gospel. You do not have to question everything, as it may be offensive. If your message is good, they will listen. They know you are a woman.' This message is frequently conveyed by male pastors and veteran female preaching voices.

While this adage may seem good advice to the new female preaching voice, it wreaks of conforming and believing that the culture will change without struggle. Those who seek the truth, to tell the truth, must be willing to sacrifice popularity for prophetic preaching. The cost of speaking truth to power is reflected in a post written by the Black feminist Rebecca Walker.

"When people ask what I would tell my younger self, the budding writer at the beginning of her career, it is always the same: I wish I could have prepared myself for what happens to a writer when she is brutally honest when she speaks truth to power in a raw and emotional way. The literary establishment continues to privilege work that's just a touch removed, 'refined' they would call it. Writers who tone down their anguish, rage, and non-traditional, 'deviant' choices are perceived as more skilled and worthy of critical acclaim. This often has a lot to do with racism and sexism and the stories we are 'allowed' to tell. I wish I had known all this, not because I would have done things differently, but because I would not have been so surprised by some of the dismissive responses to my work. I would have been more prepared."

Various texts enable prophetic preaching messages of repentance regarding child sexual abuse, and most have not been sufficiently addressed. Instead, they have been ignored or mentioned as secondary or consequential illustrations to promote the Godly heroes without regard to the sin committed. This book is a call to intentionally endeavor to engage in preaching gender equality and inclusivity.

Where have all the prophets gone that would give voice to CSA in a contemporary context? Is there a word from the Lord when our little Black girls are left desolate as Tamar? Does preaching always have to be celebratory? Lest we forget, 2 Timothy 3:16 (NIV) says, *"All Scripture is God-breathed and is useful for teaching, rebuking, correcting and training in righteousness."* Black Preaching must be willing to break the silence surrounding CSA.

*"Failure to address and approach taboo issues through the prophetic gift of preaching is an affront to God, and it leaves little hope of healing for survivors of child sexual abuse."*

# CHAPTER 5

# Creating Awareness of Child Sexual Abuse

In writing this book, one of the first goals was to assess the awareness of the child sexual abuse pandemic in the Black Church and community. Therefore, I began to engage family members, church members, and the Black community in conversation about the incidence of CSA. Not surprisingly, there were no conversations where persons reported they did not have knowledge of someone close to them being sexually abused as a child. Not ONE.

Many of those I spoke with were willing to provide their personal stories. However, when asked about their church's response to their trauma of CSA, the key person mentioned was their pastors, and the responses followed a similar pattern. Excerpts of a few responses are as follows:

A Black female devout church member in her 30s:

*"When I discovered that my 11-year-old daughter was receiving inappropriate sex messages via text from my husband, I was devastated. I immediately moved out of my house. When I told my Pastor, he said he would pray for us, but I could tell that it was uncomfortable for him. I think he thought it was too political for him to get involved."*

Black male raised in the Black Church in his 30s:

*"I was molested starting at around 9 or 10 at a church in Memphis. I moved away because no one wanted to talk about it and the person that molested me still leads in the church. I couldn't tell my pastor as he was friends with him. Black Pastors will never call out their friends no matter what."*

Black female, a product of the Black Church in her 20s:

*"I was raised in the Black Church in Memphis, and all of my family attends the church where my father used to lead. I didn't find the nerve to report my years of sexual abuse until I moved away from Memphis. I didn't tell my pastor as he was friends with my dad and all my family. I was too ashamed to be around my family. I still don't know when the sexual abuse started, but it lasted throughout my teen years."*

Black female, a devout church leader in her 50s:

*"I was molested by a family member and a church member who both attended a Baptist church. Because of the negative culture toward little "fast-tailed black girls," I would not dare report that I had been molested. However, I saw the way they treated some of the little girls who were labeled after they told the pastor and youth leaders about an incident that happened to them. They were questioned about the way they dressed and acted. In other words, they were accused of tempting older men by trying to act and dress older than they were."*

Black Male Navy Chaplain, 59:

*"There are so many men that are messed up mentally from issues of being molested that you would not believe. It is*

*sad. They sleep around with as many women as they can with no commitment. I really believe they are battling with the issue of homosexuality and true manhood because they cannot publicly admit that they were molested by another man. And this is not a black or white problem; it's across the board."*

This book aimed first to engage the community and church laity on a journey of discovering truths about the silence of the Black Church and Black Church preaching about child sexual abuse and then secondly to advocate preaching as a tool to break the silence surrounding CSA in the Black Church. To assess the awareness, or lack thereof, of CSA in the Black Church, I solicited 21 people who were either part of the Black community, Black Church, or both regarding their experiences with CSA. The results were staggering.

According to the 21 responses from the Questionnaire, 18 out of 21 participants knew a family member or close friend that had experienced CSA. The Project Director shared this information with the participants. All the responses indicated they were all very aware of the high incidence of CSA.

At this point, I transitioned the conversations to the topic of R. Kelley and the issues surrounding his accusers.

The participants were asked to share their thoughts about the Black Community blaming the victims, the victim's parents, and the Hollywood industry that shielded him, enabling his years of continual CSA. Interestingly, none of the participants defaulted to the "protectionism" train of thought that is so characteristic of the Black Community. Protectionism is a term used to describe how the Black Community covers for, denies or ignores sexual misconduct by the black male.

I then solicited the laity's thoughts about the reactions of the Black Church to a *Lifetime Series* about a black musical legend and the #MeToo movement. This question was introduced because so many participants made references to it and suggested similarities between Hollywood cover-ups for high profile stars and the Black Church practice of covering for its leaders in authoritative positions.

Participants were visually uncomfortable, some angry, some with a look of disgust, but all showed concern about the issue and were very open and honest. Some excerpts from some of the comments are below:

*(Male 40's) I'm not saying I don't believe the women. It seems awfully coincidental that when Black men like Bill or Michael get too big for the White Community, something like this happens.*

*(Female 20's) I think the parents knew that the musician was a sexual predator. Why would they let their daughters be alone with him? He was already messing with young girls. That's just crazy.*

*24-year-old female) I don't understand why the mother of one of the girls is trying to speak out now. She was the one that would not let her daughter testify against him and let him get off scot-free to do it again.*

At this point in the conversation, I directed the participants to consider, "A mother will always be protective of her children. We are talking about a teenager whose picture was being blasted all over for the world to see, that a famous black icon did this despicable thing to her. I'm not sure; as a mother,

I would not protect my daughter at all costs to prevent her from the public shame and re-victimization she would likely experience from her peers and the community at large." There were more responses afterward, but all agreed that none of us knew what our actions would be and never would unless we walked in another's shoes.

We continued with a discussion around how it has often said that "the Black Church is so protective of its male leaders that it willingly sacrifices its female members." The class became immediately engrossed in conversation, calling out Scriptures they often heard religious "Black Folks" use to protect the image of Black Pastors or Church Leaders.

Psalm 105:15 (KJV) says, *"...touch not the head of my anointed and do my prophets no harm."* Isaiah 52:7 (-says, *"How beautiful on the mountains are the feet of those who bring good news..."* and *"How can they hear without someone preaching to them"* While some referred to the quoting of Scripture when someone critiqued the misconduct of the Black Preacher, others recalled older saints saying things like, "You better watch how you put your mouth on the man of God. You just must obey. It's up to God and God alone to speak to his chosen one. Do you know what happened to the children that laughed at Elisha? (2 Kings 2:23-24), *"And a bear ate them."*

All participants in the discussion seemed to be amused by how theology has been used to manipulate and control. For example, one millennial in the group reported that her peer group doesn't come to church is because of the church's hypocrisy.

Participant (Black female 20's):

*"In Black church, you don't talk about sex at all, so why would someone report if they are being sexually abused? That would make them look bad."*

Participant (Black female 50's):

*"Women themselves are more religious than men, and they are the ones that make it bad for other girls and women. They are so protective of the man that they don't care about the children. It's all to keep up the appearance of purity and holiness."*

Participant (Black male 30's)

*"The popular black musician was able to get away with what he did because of money. A lot of people were making money off his talent. It's the same with Black Pastors and Preachers. They have the "good ole- boy network" money-making machine. It's all about the dollar. I'm not saying all of them. I'm just saying they know.'*

Because of the nature of this book and my findings as I researched it, I committed to intentionally engaging in conversation with my children, grandchildren, nieces, and nephews about CSA, what it means, and who could be a perpetrator. It broke my heart when I spoke to one of the males who shared that an older neighbor touched him inappropriately when he visited my house as a child, around the age of seven. I asked him why he never shared it with me. He responded, "I don't know. Nothing happened. I was able to get away."

It is my hope and prayer that his statement that "nothing really happened" is true. I am keenly aware of the stigma associated with boys reporting sexual misconduct. I left these parting words with my laity participants, "Just remember, you may be the one that causes a perpetrator to think twice before abusing a child, or you may be the cause of a child being made aware that anyone can be the sexual abuser. These conversations need to be in the pulpit, the pews, in the community, and in our homes."

*"The reality is that Black preachers are still hesitant to confront the issue of child sexual predators and abusers forthrightly, and some still don't see it as a prevalent issue."*

# CHAPTER 6

## Pastoral Perception of Child Sexual Abuse in Church

In researching for this book, I wanted to talk with not just members of the Black community and the Black church but also leaders within the Black church. Therefore, I sent out 100 surveys with three questions to Black Pastors and Clergy Leaders. Out of the 100 sent, 55 responded. All surveys were submitted anonymously.

*Survey Question #1 Do you believe that Child Sexual Abuse should be addressed from the Pulpit*55 out of 55 responded to this question. Additionally, 47 out of 55 believed that CSA is so prevalent in our community that it should be preached.

*Survey Question #2 What text would you choose to preach a sermon about child sexual abuse?* Out of the 55 Pastors and Clergy leaders surveyed, 47 responded to this question, of which 10 suggested New Testament Scriptures from the Gospels that did not specifically address sexual abuse, 9 suggested the story of Tamar (2 Samuel 13:1-21), 3 suggested the story of Dinah (Genesis 34).

Interestingly, out of the participants who chose to suggest a Scripture, 25% of those surveyed suggested a "text of terror" that described sexual trauma. Other suggestions were: the story of Lott's Daughters (Genesis 19:30-38), Bathsheba (2

Samuel 11:1-15), the woman with the issue of blood (Luke 8 43-48), and some were non-specific, suggesting the parables of Jesus in the Gospels. This indicated to me that most pastors and clergy are fully aware of the high incidence of CSA; they have just not preached about it. While many possibilities may hinder this type of preaching, they were not readily evident in answering the third and final question.

*Survey Question #3 In your opinion: Does the fear of discovery hinder Black Pulpit Preaching regarding child sexual abuse?*

This question was a three-part question that requested respondents to identify their gender as well as provide feedback regarding specific fears that might discourage preaching about CSA. Out of 55 respondents, 47 answered this question. One respondent stated, "I am a female clergy person. I am not sure what you mean by "fear of discovery." The subject is not off-limits in our pulpit. However, it may be uncomfortable for male preachers. The factors that would contribute to silence, in my opinion, would be ignorant of how to discuss it, guilt related to one's own past or present, or if one is aware of a situation in the congregation and has failed to address it."

As was made clear during laity sessions, so it was to the Pastors and Clergy that this was not a lynching of Black men and Black Pastors. We were not there to speak about adults and their sexual misconduct. Instead, we were there with a specific focus of struggling together as a church body to express concern and give voice to the voiceless, specifically our children being sexually abused in our homes and churches. Therefore, in addition to conducting the surveys, I interviewed four Pastors and Clergy at their respective churches. After sharing the responses from the surveys, I asked them if they had any child sexual abuse prevention measures in place. Out of the four, none were specific. None had a partnership with Memphis Child Advocacy (MCA). One reported attending a few sessions with MCA. All agreed that their experiences with CSA were reactionary and not precautionary. They all were a little uneasy with the line of questioning yet very cooperative.

Next, I asked them if there had been any reports of CSA in their church or by any of their members? All Pastors, except one, reported having dealt with at least one report of CSA from one of their members not within the church or on church grounds. One Pastor had a leader that was charged with CSA of that leader's own daughter and stated that his church dealt with it and had outside sources like MCA to assist at that time. However, after that incident, he stated, "I'm embarrassed to say that we haven't taken any other actions since then." This indicated that CSA is mostly dealt with in a reactionary response versus a precautionary. This is a problem as we consider the Pastors role as shepherd and protector of the sheep.

Another Pastor stated that she didn't think that CSA specifically should be addressed from the pulpit and that they had Youth Sessions in place that addressed this type of issue. I asked, "How does that address the perpetrator and would-be pervert that is sitting in your pews, directing your choir, or preaching in the pulpit on a Sunday morning?" I assured her that I was not trying to target her specifically; I was just seeking an earnest response. She assured me that she thought that her members felt comfortable enough to share with her if they had an issue with CSA. At this point, the Pastors were asked to respond to the following statement:

The uncomfortable truth that has long existed in the African American Community is that black girls are often on their own in fighting abuse and misogynoir. Here are a few responses:

*Female Pastor:*

"I agree that this is a problem. I have mentioned it in many sermons. I admit I have not preached an entire sermon that focuses on CSA. I also have a friend who is male, and he has shared his story. I think that talking about it from the pulpit may re-traumatize the victim. I'm not certain. But I agree we need to identify better ways the church can address it."

*Male Pastor:*

"While I am aware of the "MeToo" movement and reporting and believe there is some legitimacy, I'm not sure the Black Church has as big of a problem with this issue as other churches. We must also be careful that we don't send people chasing down a rabbit hole with accusations. This could make it more difficult for men in leadership to be comfortable with women in certain positions."

*Female Pastor:*

"That most certainly is true, and I can attest to the misogynoir as a female Pastor. We must do better in addressing this issue. We have Youth Counselors and a Youth Ministry available to all our youth. If there are a lot of reports of CSA, I am not aware. I am not saying that it doesn't happen. I am saying we don't have a lot of reports about it."

*Male Pastor:*

"That is the truth, I am sorry to say. It is a cultural part of our community. And, I am willing to participate in any way that is suggested if it will help the problem."

These responses highlight how the African American Community has struggled to deal with the abuse of black girls and women. Unfortunately, addressing this widespread problem ultimately means singling out black men, which many are hesitant to do because they don't want to become another vehicle contributing to their destruction. Based on conversations, surveys, and questionnaires, 100% of women and 80% of men would be receptive to hearing a sermon preached about CSA. However, they have never heard a sermon entirely devoted to this issue.

Based on some of the responses from the laity sessions, it is clear there is support for preaching CSA in the Black Pulpit. I'd like to highlight one response from a male participant in his 30s.

This is a gap that needs to be addressed. Black Pastors and Clergy have a voice, but why is it not utilized?

After the interviews and discussions with the Black Pastors and Clergy, I challenged them to consider preaching a sermon on CSA in the Black Pulpit.

I requested that they provide me with a time to discuss the possibility of their writing and submitting a CSA sermon that could be preached in their individual context. I proffered to each of them that this should be considered a prophetic preaching call to preach what Dr. Frank Thomas would call a "Dangerous Sermon."

In my fifty-plus years of being a child of the Black Baptist Church, I have never heard a sermon addressing this issue. In searching for a model to send to the Pastoral participants, I searched the internet for Black Preaching on this topic. Finally, I discovered a sermon titled "Surviving Lot" (Genesis 19:30-38) by Dr. Howard-John Wesley. This "Dangerous Sermon" was emailed to the participants to serve as a model for their sermon.

None were comfortable choosing the Lot text for said purpose, so I suggested that they choose whatever text they thought would be appropriate for their context, not necessarily comfortable for their context.

The interviews and discussions with the laity and the Pastoral and Clergy leaders were very informative. They really captured the helplessness that the laity feels to be able to make a difference in advocating for our children. All participants agreed that silence was a cultural norm. It was probably easier for the child to say nothing than being revictimized by the saintliness and purity of religious church antics. Some religious practices, like calling out the teen mom to come to the front of the church and repent of her sin when the teen boy or whomever (maybe even the sexual predator) was nowhere on the scene, contribute to the silence about CSA in the church. This celebration of purity practices is done without regard to the child or teen who may have been the victim of CSA. The preaching in the pulpit that "Children obey your parents" and respect and honor your elders, with no regard to preaching about sexual abuse to the "elder predators" and its prevalence in the Black community, also contributes to victims' silence. These sessions with the laity created a safe space for participants to convey their frustration with the Black Church and its silence on CSA. Participants were able to share their personal experiences anonymously, but I was still able to convey these experiences and thoughts to Pastoral and Clergy Leaders. It confirmed the importance of Black Pastors and Black Preaching in the community and challenged Black Pastors to explore preaching about CSA in the pulpit.

# CHAPTER 7

# Harsh Reality of Child Sexual Abuse in the Black Church

The research for this book was personally challenging yet rewarding. The many interviews, sessions, and surveys conducted were enlightening and alarming. It was grueling at times to sit with the responses of my male clergy participants without being judgmental. However, I was constantly reminded not to allow my learned biases and prejudices to disrupt or compromise the integrity of the discussions and research.

The reality is that Black preachers are still hesitant to confront the issue of child sexual predators and abusers forthrightly, and some still don't see it as a prevalent issue.

Especially when there were so many conversations with CSA survivors and family members of CSA victims that could be lifted in any Black Church to debunk the Black Pastoral notion that, "It is not happening to that many members at my church, or they would feel free to confide in me."

For the Pastors and Clergy that participated in this journey, it makes me hopeful. While the pastors did not specifically confirm they would preach a sermon with a focus on child sexual abuse, they were gracious to participate in this project and submit a sermon to me for research purposes.

As I reflect on interviews with Black Pastors, I am not sure that denial is the driving force behind their silence. It is seemingly more of fear to call out some of the members with which they are in a relationship. It is most difficult to admit the heavy burden of shame that has been thrust upon the victims and families of CSA. And a necessary part of righting this wrong is for Black Preaching to sound the alarm, make the clarion call from the pulpit that CSA and the culture that helped create it must CHANGE NOW.

This was my basic feeling throughout conversations, interviews, class sessions, reviewing questionnaires and survey responses, and reading sermons from clergy participants. Unfortunately, while conducting this project, it became increasingly obvious to me that it is most certain that many pockets of ministry continue to promote a culture that seeks to keep women and girls ignorant about what is really said in the Bible about sexual abuse. While I am very much aware of the impact that this rape culture can have on our Black boys and young men, I am hopeful that pastors will begin having conversations in their networks that encourage a gender ethic of respect, equality, and appreciation for all so that our little Black boys don't grow up to be sexual predators.

# CHAPTER 8

# Reflection & Time for Change

As I reflect on the past journey to complete this project as an African American female preacher, I was struck by these words. "Words are everywhere, inside me, outside me...I'm in words, made of words, other's words...I'm all these words, all these strangers, this dust of words, with no grounds for their settling, no sky for their dispersing, coming together to say, fleeing one another to say, that I am they, all of them." (McClure. Kindle file) This is how I feel about Black Preaching and CSA.

I am constantly being reminded of the foolishness of preaching as I navigate through this journey called life, on this place called earth, that was spoken into existence by yes words. I am overwhelmed by the reality that it is the word spoken and the one unspoken that has created our present reality. If this project and seminary experience has confirmed anything, it is this "Death and life are in the power of the tongue." (Proverbs 18:21 NKJV) Our words matter.

This leads me to reckon with questions about my own preaching as it relates to being prophetic in this vitriolic season of the attack from the enemy on our little ones. I cannot help but entertain the thought that our "Patriarchy Preaching Predecessors" share a grave responsibility in exacting a burden that challenges Black Pastors to distance themselves from their inherited patriarchal culture.

Old Testament prophets had the burden of defying their normative culture to hear the word of God. They were heard and were obedient in proclamation to the masses that did not want to hear it. Down through the years, I have heard Amos's words from the pulpit, but now they speak so much differently to me and through me. The gravity of these words was not only astounding but somewhat burdensome to my spirit. Are we becoming a nation whose time is ripe for judgment?

As I attempted to interpret the information and testimonies for this book, I remember feeling the discomfort that it was sure to cause my brothers in ministry. Yet, the silence must be broken with the many unanswered questions that loom in the wilderness of discomfort and shame to the victim. While looking into the eyes of one of the participants in the class, there was a visceral feeling that she had been the victim of CSA.

It has been noted in many conversations with married women that they never shared their CSA experience with their husbands. Many expressed the shame that it would cause to their families and the burden of responsibility it would put on them to be the voice for others, not realizing that the sexual predator has a 100% chance of abusing more victims. The sad reality is that this will not stop anytime soon. While sharing a snippet of my story of CSA, one of the participants stated: "Rev. Stephanie, everybody is not as strong as you, and they will not come forward knowing the wrath of church folks about exposing family secrets." These and many other voices that express shame and fear continue to disturb comfortable preaching and teaching.

I remember wanting to ask the question of the preacher participants: Is there a prophetic preacher in the house? Where have all the prophets gone? Would our Spiritual Leaders today have the courage to admit that something needs to change, and it needs to change now?

According to McMickle, "The road back to authentic prophetic preaching in America's pulpits may begin with the realization that at the turn of the 21st century, a remnant is all that remains in both the pulpit and the pew. Today with so many preachers focused on matters other than prophetic preaching, there is an urgent need for preachers who will continue to pursue the agenda of justice and righteousness. A remnant may be all that is left, but a faithful remnant in the hands of a just and loving God can make a big difference in our world." (McMickle. Kindle file)

*"Today with so many preachers focused on matters other than prophetic preaching, there is an urgent need for preachers who will continue to pursue the agenda of justice and righteousness."*

# CHAPTER 9

## Personal Testimony

*We contend that there is another kind of justice, restorative justice, which has the characteristic of traditional African jurisprudence. Here the central concern is the healing of breaches, the redressing of imbalances, the restoration of broken relationships, a seeking to rehabilitate both the victim and the perpetrator, who should be given the opportunity to be reintegrated into the community he "or she" has injured by his offense.*

Bishop Desmond Tutu

As a survivor of child sexual abuse and having had a near-death experience at puberty, I am of the opinion of the apostle, Paul. "I have become all things to all child sexual abuse victims and survivors that I may by any means save one." The impact of CSA in my life in the past caused much anguish and had the capacity to cause me to self-destruct.

There were times as a teen and young adult I was appalled by the hypocrisy and sexist culture of the Black Church. While I never remember being angry, I do remember I internalized the feelings of never being able to measure up to the erroneous picture of perfection seemingly personified in Black Church rhetoric. The constant reminders of purity and virginity from the pulpit without the mention of sexual predators that raped the body and choice of a child was deafening. Then, there was the participatory revictimization from the congregation to call a teen girl to the front of the church, adding to the burden of silence, a public burden of shame.

Like Tamar, I remember bringing my shame to the church and feeling re-victimized for telling my story. Around the mid-1980s, at the age of 21, not yet claiming CSA survivor status, I began to feel the call of God on my life to tell my story so that others may have the courage to do the same. As a teen mom, my vision of being a doctor was blurred, so I chose to pursue a career as a registered nurse. This would be less demanding and wouldn't require as much time. Yet, I felt led to study the word of God more intensely through seminary training. I was convinced that seminary was my source of hope to prepare me for a teaching ministry. But, the thought of being a Preacher never entered my mind then. Why would a black teenage mother, having never been exposed to a female preaching voice, feel "the call" to preach? Perhaps seminary training was an opportunity to minister to others, thereby ministering to myself.

I wrote a long letter with my application conveying my journey from CSA to my current state as a teen mom. I'm not sure what I expected but certainly not the content of the rejection letter. I remember the exact words, "We have reviewed your application, and we regret to inform you that you do not meet the criteria representative of our Institution."

This broke my heart, and I regretted sharing my burden of shame and hope for the future in such intimate detail. But, like the many times that I would feel the condemning eyes of the church as a teen mom, I wanted to shout to the world. You don't understand my choice was taken away from me! Yet, I learned to distance myself from feeling anything. I adopted a spirit of survival that canceled any time for unwanted emotion. The rejection from this Christian institution reaffirmed my shame and sealed my fate of never speaking publicly or otherwise about my molestation. I stand truly amazed that God allowed me to more than survive the trauma of CSA, despite the church's antics of revictimizing the victim.

Tamar's story of sexual trauma remains as riveting as the depiction of black culture in the fiction-based film *The Color Purple*. The Bible says that Tamar was left desolate after she was raped, and so did Celie. "When Shug Avery asked Celie if she ever got mad, she stated, "I can't even remember the last time I felt mad, I say. I used to git mad at my mammy cause she put a lot of work on me. Then I see how sick she is. Couldn't stay mad at her. Couldn't be mad at my daddy cause he my daddy. Bible says, Honor father and mother no matter what. Then after while, every time I got mad or start to feel mad, I got sick. Felt like throwing up. Terrible feeling. Then I start to feel nothing at all." Celie had been forced to stay in silence for so long that she was not even able to get mad or angry about what had transpired in her life because of her loyalty to God.

"The excesses of Celie's sufferings, however, fit into a narrower literary pattern than simply that of this fictitious characterization. It resonates within a literary context to provide not so much an attack on Black males as an examination of the very nature of women's passivity and women's defenses. Celie's acquiescence is neither extreme in its individuality nor socially threatening. Instead, it is the covert resistance of a woman, forced like Tamar, to fit into an alien world and make it her own." This is our present context and reality in the Black Church for little Black girls who are being abused and for those who are survivors of CSA.

("*The Abuse In The Color Purple English Literature Essay.*" UKEssays.com. 11 2018. All Answers Ltd. 03 2019 (https://www.ukessays.com/essays/english-literature/the-abuse-in-the-color-purple-english-literature-essay.php?vref=1 )

CSA is such a shameful and traumatizing issue for so many reasons other than just the act itself. Given that sexual trauma against our little girls has painful biblical origins, it would make sense that preaching should consideration this. If most sermons heard by little girls are preached, heralding patriarchal and misogynistic figures, what message does this send to girls and women who know about Tamar, Bathsheba, and the many women who these men of God have abused?

Black Preaching attempts at preaching a liberation message as all-inclusive and gender-neutral have falsely relieved the preacher's responsibility to address CSA specifically address.

There is no greater force than the Black Church that has the power to lift a spirit and destroy it in the same moment. None greater that can teach liberation with the same zeal as it exacts oppression. None greater to exemplify the unconditional love of Christ yet render others unlovable using the same biblical text. The Black Church has the sway to point out your uniqueness and giftedness while preaching you into shame and silence.

This is the reality for victims and survivors of CSA in the Black Church and community. During interviews with pastors and listening to their responses to this endemic, it is evident that intentionality in preaching this subject had not been on their preaching list. Or rather, it was not something they were comfortable preaching as a primary focus. There cannot be an argument for preaching relevancy with the interviews, research, surveys, and questionnaires. This being noted, there must be an intentional focus to adequately and rightly divide the word of truth related to CSA.

There have been numerous studies reporting that most sexual abusers have been victims of CSA. It has also been noted that counseling has not been an accepted norm for CSA victims within the black community for many reasons. Given the high incidence of mental illness and suicide, one must consider the voice of the Pastor as vital to the well-being of the Black Church and the community. Our silence is unacceptable and should be called out by every victim that has kept silent to protect the name of their families, friends, coaches, doctors, teachers, preachers, or pastors.

The primary goal of the Black Church should be to ensure that our little Black girls and boys disclosing sexual abuse are not further victimized by the Church that God established and the preachers that God sent to speak for them and protect them. If this book carries any weight, it is my earnest desire that it is the burden of the Black Pastor to consider the audience of the one in four victims present, the predators in the pews and the pulpit, and the silent enablers in church and community when considering what is relevant in this Preaching Season. The voice of the Pastor or Preacher can and does make a difference.

As I reflect on the totality of this book and all that has transpired, I recall the Revelation given to John the Apostle from "The Message" Bible's contemporary interpretation. "Write what you see into a Book, and send it to the Seven Churches." (Revelation 2:1-7)

*Revelation to The Black Churches from a Survivor of Child Sexual Abuse Making Plain What is Necessary to Happen*

*(A redacted and edited version)*

*I, the little Black girl who has grown up in the trenches of your patriarchal black church culture: All the best to you from the God I love and serve, and from the Spirits of my sisters Tamar, Dinah, and Lot's daughter, and from Jesus Christ our Lord and Savior who has declared that all are equal in the sight of God.*

*I am writing what I see and hear to the churches, which should be aware that 1-4 little girls sitting in your pews on Sunday morning are victims of child sexual abuse.*

*To the Black Church in Memphis. The Pastor of the Church. The one with the thriving ministry with fancy sanctuaries, luxury cars, fine houses, and golden status within the church, I speak:*

*"I see what you've done, your hard work, your refusal to quit. I know you shun evil language and the appearance of sin. I know you preach fiery sermons on Sunday and holy-filled Bible studies on Wednesday nights. But you walked away from the little ones that needed you the most. In your zeal to be the best and have the best and fellowship with the best, your love for the little ones waxed cold. They were right in front of you, sitting in your pews, attending your vocational Bible school classes, serving on the junior usher board, singing in the sunshine choir, and reading Scripture on your children and youth days."*

*What is going on with you? Do you have any idea that the children are counting on you and they are watching you to see if you are a messenger from God and not a cover for the sexual predator?*

*Black Preacher, turn back! Recover the innocence and the faith of our children!*

*You do have this to your credit; You hate the evil of racism. I, as a black woman, hate it too.*

*Are your ears awake? Listen. Listen to the pain, anguish, and cries of the victims, the pain and anguish of the survivors, and be alert to the perpetrators in your midst. The spirit of repentance from years of silence is blowing through Black Church. God is watching along with the little ones to see how you respond.*

*Black Church Black Preacher Black Teacher Black Leaders, Black families, Black Sisters, Black Brothers. A message from the servant of God who is a survivor of CSA...DO BETTER, BLACK CHURCH.*

*"If most sermons heard by little girls are preached, heralding patriarchal and misogynistic figures, what message does this send to girls and women who know about Tamar, Bathsheba, and the many women who these men of God have abused?"*

# CHAPTER 10

# Final Words

The Black Church and the Black family are the two most important institutions within the Black community (Fulton, 2011). Lincoln and Mamiya (1990) defined the Black Church as the Black-controlled independent denominations that make up the heart of Black Christianity in the United States. According to the 2007 U.S. Religious Landscape Survey conducted by the Pew Research Center's Forum on Religion and Public Life, 87% of African Americans describe themselves as belonging to one religious group or another (Neha Sahgal and Smith, 2009).

As such, the Black Church is in a strategic position to be useful in ameliorating a plethora of social problems and conditions that plague the African American community. Despite the significant, influential position of the African American church, it has failed its members in its response to many critical social issues." And one of those issues is CSA. (Moore, Cassie. 2012. Print)

Finally, the issues contributing to child sexual abuse are as complex and varied as the journey to prevention and eradication. It is my opinion of an unknown author's perspective of ministry "For myself, I find I become less cynical rather than more—remembering my own sins and follies; I realize that people's hearts are not often as bad as their acts, and very seldom as bad as their words."

I will continue to challenge the heart of Black Pastors and Preachers to postulate moral imagination in their preaching practice and theology to advocate for victims and survivors of child sexual abuse.

# Bibliography

Allen, O. Wesley. *Preaching and the Human Condition: Loving God, Self, & Others*. Nashville:

Abingdon, 2016. Print.

Ashmore, Malcolm & Reed, Darren (2000). *Innocence and Nostalgia in Conversation Analysis: The Dynamic Relations of Tape and Transcript*. Forum Qualitative Sozialforschung / Forum: Qualitative Social Research, 1(3), Art. 3.

Baldwin, Joyce G..1 and 2 Samuel (Tyndale Old Testament Commentaries) (p.264). InterVarsity Press. Kindle Edition.

Bartlett, Linda D. *The Failure of Sex Education in the Church: Mistaken Identity, Compromised Purity: Questions & Answers for Christian Dialogue*. Iowa Falls, IA: Titus 2 for Life, 2014. Print.

Cameron, Helen, John Reader, and Victoria Slater. *Theological Reflection for Human Flourishing: Pastoral Practice and Public Theology*. London: SCM, 2012. Print.

Clark, Kurt S. *"Invisible Institution and Empire: A Culture of Resistance in America." Barnes & Noble*. Library of Congress North Charleston, South Carolina, 15 Dec. 1359. Web. 03 Nov. 2017.

Cook, Guy (1990). Transcribing Infinity: Problems of Context Presentation. *Journal of Pragmatics*, 14, 1-24.

Dickhaut, Walter R. *Building a Community of Interpreters: Readers and Hearers as Interpreters*. Eugene, OR: Cascade, 2013. Print.

Ekblad, Bob. *Reading the Bible with the Damned*. Louisville, KY: Westminster John Knox, 2005. Print.

Florence, Anna Carter. *Preaching as Testimony*. Louisville, Ky: Westminster John Knox, 2007. Print.

Foskett, Mary F. *Interpreting the Bible: Approaching the Text in Preparation for Preaching*. Minneapolis, MN: Fortress, 2009. Print.

Frawley-O'Dea, Mary Gail. *Perversion of Power: Sexual Abuse in the Catholic Church*. Nashville (Tenn.): Vanderbilt UP, 2007. Print.

Goldner, Virginia, and Mary Gail Frawley-O'Dea. *The Sexual-abuse Crisis and the Catholic Church*. Lawrence, Kan.: Analytic, 2004. Print.

Graham, Elaine, Heather Walton, and Frances Ward. *Theological Reflection: Methods*. London: SCM, 2005. Print.

Have, Paul ten (1990). Methodological Issues in Conversation Analysis. *Bulletin de Méthodologie Sociologique,* 27 (June), 23-51."

Heimlich, Janet. *Breaking Their Will: Shedding Light on Religious Child Maltreatment*. Amherst, NY: Prometheus, 2011. Print.

Hendricks, Obery M. *The Politics of Jesus: Rediscovering the True Revolutionary Nature of the Teachings of Jesus and How They Have Been Corrupted*. New York: Doubleday, 2006. Print.

Hendricks, Obery M. *The Universe Bends toward Justice: Radical Reflections on the Bible, the Church, and the Body Politic.* Maryknoll, NY: Orbis, 2011. Print.

Jacobsen, David Schnasa., and Robert Allen. Kelly. *Kairos Preaching: Speaking Gospel to the Situation.* Minneapolis, Minn: Fortress, 2009. Print.

Kearney, R. Timothy. *Caring for Sexually Abused Children: A Handbook for Families & Churches.* Downers Grove, IL: InterVarsity, 2001. Print.

Kent, Brianna Black. *The Process of Healing for Adult Male Survivors of Childhood Sexual Abuse by Catholic Priests.* N.p.: n.p., 2006. Print.

Lehman, Carolyn. *Strong at the Heart: How It Feels to Heal from Sexual Abuse.* Arcata, CA: Sky Pilot, 2014. Print.

Lose, David J. *Preaching at the Close of Age Preaching at the Crossroads.* Lanham: Fortress, 2013. Print.

McManners, Peter. *Corporate Strategy in the Age of Responsibility.* Farnham: Gower Limited, 2014. Print.

McMickle, Marvin. *Where Have All the Prophets Gone?: Reclaiming Prophetic Preaching in America.* Cleveland, OH: Pilgrim, 2006. Print.

Moore, Cassie. *Did You Hear Me Crying: The Moving Story of Survival through 45 Years of Sexual, Physical, and Emotional Abuse.* London, United Kingdom: Live It, 2012. Print.

Moore, Sharon & Robinson, Michael & Dailey, Alicia & Thompson, Carlos. (2015). Suffering in Silence: Child Sexual Molestation and the Black Church: If God Don't Help Me

Who Can I Turn To? *Journal of Human Behavior in the Social Environment*. 25. 10.1080/10911359.2014.956962.

Oubre, Angela. *Exploring Emotional Intimacy Among African American Female Survivors of Childhood Sexual Abuse Who Utilize Black Church Support Services*. South Orange, NJ: Seton Hall U, 2004. Print.

Rainer, Thom S. *Who Moved My Pulpit?: Leading Change in the Church*. Nashville, TN: B & H Group, 2016. Print.

Rthangel73. "Eddie Long And The Black Church's Legacy Of Child Sexual Abuse." *News One*. News One, 17 Jan. 2017. Web. 03 Nov. 2017.

Salter, Anna C. Predators (p. 11). Basic Books. Kindle Edition

Stewart, Warren H. *Interpreting God's Word in Black Preaching*. Valley Forge, PA: Judson, 2001. Print.

Stone, Howard W., and James O. Duke. *How to Think Theologically*. Minneapolis: Fortress, 2013. Print.

Tisdale, Leonora Tubbs. *Prophetic Preaching: A Pastoral Approach*. Louisville, KY: Westminster John Knox, 2010. Print.

Volf, Miroslav. *Captive to the Word of God: Engaging the Scriptures for Contemporary Theological Reflection*. Grand Rapids, MI: William B. Eerdmans Pub., 2010. Print.

*Walk in the Light: A Pastoral Response to Child Sexual Abuse*. Washington, D.C.: United States Catholic Conference, 1995. Print.

Walker, Alice. *Overcoming Speechlessness: A Poet Encounters the Horror in Rwanda, Eastern Congo, and Palestine/Israel*. New York: Seven Stories, 2010. Print.

Online:

Gerald O. West, "Tamar (2 Samuel)", n.p. [cited 4 Apr 2019] https://www.bibleodyssey.org:443/en/people/mainarticles/tamarhttps://religionnews.com/2014/01/09/startling-statistics/

https://www.apa.org/pi/about/newsletter/2014/11/child-sexual-abuse

https://sojo.net/articles/sexual-violence-and-church/i-am-pastor-and-rape-survivor-metoo-opportunity-church

www.rolereboot.org/culture-and-politics/details/2014-09-sexual-abuse-code-silence-black-community/
(Cherise Harwell)

https://racebaitr.com/2015/12/24/hollering-in-the-sanctuary-can-you-hear-me/
(Sevonna Brown)

https://extension.tennessee.edu/eesd/Documents/HR/Minors/TennesseeLawonMandatoryReportingofChildandChildSexAbuse.pdf

# About the Author

Rev. Dr. Stephanie Michelle Myers is an ordained Baptist Preacher and Child of the Church. She is a product of Memphis City Schools, with Mississippi roots from spending summers with her grandparents. She has served in various positions of Ministry to include, Trustee, Associate Minister, Director of Christian Education & Director of Women's Ministry.

Rev. Stephanie has been educated in numerous Schools and fields of Study to include State Technical Institute /Memphis with a major in Computer Programing, University of Memphis with a major in Jazz Vocal, Baptist School of Nursing, University Tennessee College of Health Sciences. She is also a 2009 Graduate of Delta Leadership Institute, and a 2015 Graduate of Harvard University Executive Academy Leadership School. She holds numerous designations and certifications as a Registered Nurse with 20 + years of experience. She is a 2014 Master of Divinity Graduate and 2019 Doctor of Ministry Graduate from Memphis Theological Seminary.

She is currently CEO and founder of Purple Reign Realty LLC which launched on June 9th, 2018. She has served in the Real Estate Industry since 1997 as a Licensed Broker, Certified Credit Counselor, and CDPE (Certified Distressed Property Expert). She is a member of (NAR) National Association of Realtors, (NAREB) National Association of Real Estate Brokers, (TAR) Tennessee Association of Realtors, (MAAR) Memphis Area Association of Realtors, and (NWMAR)

Northwest Mississippi Association of Realtors. She has served in various leadership capacities to include; Governmental Affairs Committee Member, Realtor Political Action Committee Member, 2nd Vice President for NAREB (Memphis Chapter) 2006-2008, and Education Liaison for NAREB 2005-2007. 1st Vice President of WCNAREB Memphis Chapter 2022.

Rev. Stephanie believes that her greatest gifts in life are her family and friends. She is the daughter of Ms. Shirley Black-Myers, proud mother, Nana, sister, aunt, cousin and friend to many.